# WHAT IS QUANTUM PHYSICS?

*How Quantum Physics Phenomena Influence Your World In an Easy and Intuitive Way*

JACOB ORSON

© Copyright 2020 by

**All rights reserved.**

This document is geared towards providing exact and reliable information with regard to the topic and issue covered. The publication is sold with the idea that the publisher is not required to render accounting, officially permitted or otherwise qualified services. If advice is necessary, legal or professional, a practiced individual in the profession should be ordered.

From a Declaration of Principles which was accepted and approved equally by a Committee of the American Bar Association and a Committee of Publishers and Associations.

In no way is it legal to reproduce, duplicate, or transmit any part of this document in either electronic means or in printed format. Recording of this publication is strictly prohibited, and any storage of this document is not allowed unless with written permission from the publisher. All rights reserved.

The information provided herein is stated to be truthful and consistent, in that any liability, in terms of inattention or otherwise, by any usage or abuse of any policies, processes, or directions contained within is the solitary and utter responsibility of the recipient reader. Under no circumstances will any legal responsibility or blame be held against the publisher for any reparation, damages, or monetary loss due to the information herein, either directly or indirectly.

Respective authors own all copyrights not held by the publisher.

The information herein is offered for informational purposes solely and is universal as so. The presentation of the information is without a contract or any type of guarantee assurance.

The trademarks that are used are without any consent, and the publication of the trademark is without permission or backing by the trademark owner. All trademarks and brands within this book are for clarifying purposes only and are owned by the owners themselves, not affiliated with this document.

# Table of Contents

**Chapter One: Introduction to Quantum Physics .. 1**

    The Development of Quantum Theory ................ 5

    The Copenhagen Interpretation and the Many-Worlds Theory ...................................................... 6

    Quantum Theory's Influence ............................... 8

    Likelihood versus Causality In Quantum Physics ............................................................................ 13

    Three progressive standards ............................. 22

    Quantized properties? ...................................... 24

    Particles of light? ............................................. 27

    Rushes of the issue? ......................................... 29

    The vulnerability standard ............................... 31

    Forward ............................................................ 32

**Chapter Two: How Max Planck Developed the New Concepts ...................................................... 34**

    The mysterious entropy .................................... 35

    Dark body radiation ......................................... 39

    Error with physics ............................................ 41

A traditionalist progressive .......................... 46
Einstein: the genuine author of quantum physics?
............................................................................ 49

**Chapter Three: The Laws that Govern Quantum Physics** ............................................................... 54

Photons: the quantization of light ..................... 56
The photoelectric impact ................................... 58
Results of light being quantized ........................ 60
The quantization of issue: the Bohr model of the molecule ............................................................ 62
Quantum electrodynamics ................................. 63

**Chapter Four: Numerous Experiments on Quantum Physics** .................................................... 66

Single-Molecule Interference ............................ 67
Quantum Non-Locality ...................................... 69
Exactness Measurement ..................................... 73
The kilogram goes quantum ............................... 74
Reality broke a little .......................................... 75
Snare got its excitement shot. ........................... 76
Warmth crossed the vacuum. ............................. 77
Quantum burrowing broke ................................. 78
Metallic hydrogen may have shown up on Earth.
............................................................................ 79
We viewed the quantum turtle. .......................... 80

A minuscule quantum physics traveled back in time. ................................................................. 81

**Chapter Five: Explanation of the Laws of Attraction and Relativity** ............................................................. 84

What Is The Law Of Attraction? ....................... 84

The History of the Law of Attraction ................ 86

Is The Law Of Attraction Real? ......................... 87

The Science Behind The Law Of Attraction: Fact, Not Fiction ......................................................... 89

The most effective method to Use the Law of Attraction ........................................................... 90

What are the 7 Laws of Attraction? ................... 91

Think about the Law of Attraction as gravity. ... 91

How And Why Does The Law Of Attraction Work? ................................................................. 93

Your Relationship to the Universe .................... 94

The Law of Attraction works similarly .............. 95

It's Not Magic, and It's Science ......................... 95

How To Use The Law Of Attraction (And What For)? ................................................................. 100

What Principles Shall I Observe When It Comes To The Law Of Attraction? .............................. 101

Law Of Relativity ............................................. 105

For what reason is it Special? .......................... 106

Fabricate your Time Machine .......................... 107

**Chapter Six: Quantum Physics in Relation with our Body and Health** ............................................. 111

    Ways That Quantum Technology Could Transform Health Care ...................... 113

    1: Improved Disease Screening And Treatment ............................................................. 113

    2: No More Needles ............................................ 115

    3: Hacking Human Biology ............................... 116

    4: More Secure Health Data .............................. 117

**Chapter Seven: Time in Quantum Physics** ........ 119

**Chapter Eight: Quantum Physics Applied to the Real World** .............................................................. 122

    Ways Quantum Physics Affects Your Daily Life ............................................................................. 128

    Toaster ovens: ..................................................... 129

    Glaring Lights: .................................................... 131

    Physics: ................................................................ 133

    Natural Compass ................................................ 135

    Semiconductor .................................................... 136

    Laser ..................................................................... 137

    Microscopy ......................................................... 137

    Global Positioning System (GPS) .................... 138

    Attractive Relativity Imaging .......................... 138

    Media transmission ........................................... 139

 Super Precise Clocks............................................ 139

 Uncrackable Codes .............................................. 141

 Super-Powerful Computers.................................. 142

 Quantum Microscope............................................ 143

**Chapter Nine: Quantum Computing** .................. 146

 What is quantum processing? ............................ 146

 How to accomplish quantum physics work? ... 148

 What can quantum physics do?.......................... 149

 When will I get a quantum physics? ................. 151

# CHAPTER ONE

# Introduction to Quantum Physics

Quantum Mechanics, mostly called Quantum Physics, is the relationship shared among energy and matter. "Quantum" is Latin for "how much." The mechanics of this alludes to a unit that quantum physics allocates as an estimation to specific physical amounts in modest quantities. Fundamentally, quantum articulations are typically taken a gander at and concentrated on a sub-nuclear level with sub-nuclear particles.

Sub-nuclear particles are little. If a molecule were as large as a house, the sub-nuclear molecule would be as large as a gumdrop inside that house's kitchen cabinet.

There were a few things that needed to occur before the investigation of Quantum Mechanics flourished. In 1838, the disclosure of cathode beams, at that point, Gustav Kirchoff, in 1850, distributed an announcement of the "dark body radiation" issue. At that point in 1877, Ludwig Boltzmann proposed that a physical framework's energy conditions could be discrete.

In 1900, Max Planck concocted a physics that energy is transmitted and retained. At that point, he delivered a recipe that would get known as "Planck's Action Constant."

Planck is as yet known as the granddad of Quantum Physics. After his physics was distributed, different researchers observed, and abruptly, you had a couple of more physics' structures until Quantum Mechanics was being speculated and concentrated all through the world.

It is a direct result of Quantum Physics: we are nearly repulsive force, having superconductors, MRI machines in emergency clinics, and now can understand time travel is conceivable.

This all sounds so phenomenal, yet this is the thing that the researchers working in the field of

Quantum Mechanics will let you know. The hardest thing for most of us to comprehend is the connection between sub-nuclear particles and the Law of Attraction.

Inside the investigation of Quantum Mechanics, it has been discovered that sub-nuclear particles take the course. Some other power is moving these structure squares of physical issues around the Universe.

After a couple of twofold visually impaired cut tests, utilizing Sub-Atomic Particles as subjects, it was found they could change from particles to wave structure at that point back once more. These particles could leave our measurement at that point pop right once more into it once more. We also discovered that these sub-nuclear particles changed from particles to wave structure, contingent upon the goal. We found that we were unable to eliminate ourselves out of the condition while testing the particles. We affected the particles by contemplating the result. There is significantly more to this.

Everything turns out to be extremely befuddling. It bewildered Einstein until the day he passed on. Getting molecule/wave duality isn't something that most of us can get our psyches around without any problem.

However, one of the speculations developed from the establishment of Quantum Physics is that we control life's very texture by contemplating it. Our contemplations have articulation that goes out. At that point brings to us what we center on. This is the Law of Attraction.

Quantum physics is the hypothetical premise of present-day material science that clarifies the nature and conduct of issues and energy on the nuclear and subatomic levels. The nature and behavior of matter and life at that level is once alluded to as quantum material science and quantum mechanics.

In 1900, physicist Max Planck introduced his quantum physics to the German Physical Society. Planck had looked to find the explanation that radiation from a gleaming body changes in shading from red, to orange, and, at long last, to blue as its temperature rises. He found that by making the presumption that energy existed in singular units similarly, that issue does, instead of a steady electromagnetic wave - as had been earlier accepted - and was like this quantifiable.

He could discover the solution to his inquiry. The presence of these units turned into the primary presumption of quantum physics.

Planck composed a numerical condition including a figure to speak to these individual units of energy, which he called quanta. The state clarified the wonder well indeed; Planck found that at certain discrete temperature levels (authentic products of an essential least worth), energy from a gleaming body will possess various shading range regions. Planck accepted a physics yet to rise out of the disclosure of quanta; however, truth be told, their very presence inferred a new and principal comprehension of the laws of nature. Planck won the Nobel Prize in Physics for his physics in 1918. However, different researchers' improvements over a thirty-year time frame all added to the quantum physics's advanced comprehension.

## The Development of Quantum Theory

• In 1900, Planck made the supposition that energy was made of individual units or quanta.

• In 1905, Albert Einstein speculated that the power, yet the radiation itself was quantized similarly.

- In 1924, Louis de Broglie suggested that there is no crucial distinction in the cosmetics and conduct of energy and matter; on the nuclear and subatomic level, either may act as though made of either particles or waves. These physics got known as the rule of wave-molecule duality: elementary particles of both energy and matter carry on, contingent upon the conditions, as either particles or waves.

- In 1927, Werner Heisenberg suggested that exact, synchronous estimation of two reciprocal qualities -, for example, the position and force of a subatomic molecule - is unimaginable. As opposed to traditional material science standards, their simultaneous estimation is inevitably imperfect; the more precisely one worth is estimated, the more defective will be to assess the other value. These physics got known as the vulnerability guideline, which incited Albert Einstein's famous remark, "God doesn't play dice."

## The Copenhagen Interpretation and the Many-Worlds Theory

The two powerful understandings of the quantum physics' suggestions for the truth are the Copenhagen translation and the many-universes belief. Niels Bohr proposed the Copenhagen translation of the quantum physics, which attests that a molecule is whatever it is estimated to be (for instance, a wave or a molecule), yet that it can't be accepted to have explicit properties, or even to exist until it is estimated. So, Bohr was stating that target reality doesn't exist. This means a rule considered superposition that guarantees that while we don't have the foggiest idea of the condition of any article, it is entirely all potential states simultaneously, as long as we don't hope to check.

To represent these physics, we can utilize the celebrated and fairly savage similarity of Schrodinger's Cat. To begin with, we have a living feline and spot it in a thick lead box. At this stage, there is no doubt that the kitten is alive. We, at that point, toss in a vial of cyanide and seal the case. We don't know whether the kitty is active or if the cyanide container has broken and the kitten has kicked the bucket. Since we don't have the foggiest idea, the cat is both dead and alive, as per quantum law - in a superposition of states. When we tear open the crate and see what condition the kitten is, the superposition is lost, and the cat must be either alive or dead.

The second translation of quantum physics is the many-universes (or multiverse physics. It holds that when an expected exists for any item to be in any express, the Universe of that article changes into a progression of equal universes equivalent to the number of potential states in which that the report can exist, with every Universe containing an attractive single conceivable condition of that object. Besides, there is a system for communication between these universes that, in one way or another, licenses all states of being open somehow or another and for all potential states to be influenced in some way. Stephen Hawking and the late Richard Feynman are among the researchers who have communicated an inclination for the many-universes physics.

## Quantum Theory's Influence

Even though researchers all through the previous century have dismissed the ramifications of quantum physics - Planck and Einstein among them - the physics' standards have consistently been upheld by experimentation, in any event, when the researchers were attempting to invalidate them. Quantum physics and Einstein's physics of relativity structure the reason for current material science. The standards of quantum material science are being applied in an expanding number of zones, including quantum optics, quantum science, quantum processing, and quantum cryptography.

Quantum mechanics is a physical science managing the conduct of issue and energy on the size of iotas and subatomic particles/waves.

It likewise shapes the new comprehension of how huge articles, for example, stars and systems, and cosmological occasions, such as the Big Bang, can be dissected and clarified.

Quantum mechanics is establishing a few related orders, including nanotechnology, dense issue material science, quantum science, basic science, molecule material science, and gadgets.

The expression "quantum mechanics" was the first begat by Max Born in 1924.

The acknowledgment by the overall material science network of quantum mechanics is because of its exact expectation of frameworks' physical conduct, including frameworks where Newtonian mechanics comes up short.

Indeed, even general relativity is restricted - in

manners, quantum mechanics isn't - for depicting frameworks at the nuclear scale or littler, at extremely low or exceptionally high energies, or the most minimal temperatures.

During experimentation and applied science, quantum mechanical physics has been demonstrated to be fruitful and functional.

The establishments of quantum mechanics date from the mid-1800s, yet the genuine beginnings of QM date from crafted by Max Planck in 1900.

Before long, Albert Einstein and Niels Bohr made significant commitments to what exactly is presently called the "old quantum physics."

Notwithstanding, it was not until 1924 that a complete picture rose with Louis de Broglie's issue wave theory and the genuine significance of quantum mechanics turned out to be exact.

The most noticeable researchers to contribute during the 1920s to what is currently called the "new quantum mechanics" or "new material science" were Max Born, Paul Dirac, Werner Heisenberg, Wolfgang Pauli, and Erwin Schrödinger.

Afterward, the field was extended with work

by Julian Schwinger, Sin-ItiroTomonaga, and Richard Feynman to advance Quantum Electrodynamics in 1947 and by Murray Gell-Mann specifically for the improvement of Quantum Chromodynamics.

The obstruction that produces shaded groups on bubbles can't be clarified by a model that portrays light as a molecule.

It very well may be clarified by a model that portrays it as a wave.

The drawing shows sine waves that look like waves on the outside of water being reflected from two surfaces of a film of changing the width, yet that portrayal of light's wave idea is just an unrefined similarity.

Early analysts contrasted in their clarifications of the basic idea of what we presently call electromagnetic radiation.

Some kept up that light and different frequencies of electromagnetic radiation are made out of particles, while others stated that electromagnetic radiation is a wave marvel.

In old-style material science, these thoughts

are commonly opposing.

Since the beginning of QM, researchers have recognized that neither thought without anyone else can clarify electromagnetic radiation.

Notwithstanding the achievement of quantum mechanics, it has some questionable components.

For instance, the conduct of minute items portrayed in quantum mechanics is altogether different from our everyday experience, which may incite some level of wariness.

The more significant part of traditional material science is perceived as exceptional quantum material science physics and relativity physics.

Dirac offered the relativity physics a powerful influence for quantum material science to appropriately manage occasions that happen at a considerable portion of the speed of light.

In any case, traditional material science also manages mass fascination (gravity). Nobody has yet had the option to carry gravity into a bound together physics with the relativized quantum physics.

## Likelihood versus Causality In Quantum Physics

It is asserted that quantum material science is not concerning conviction (for example - causality) yet on likelihood. Subsequently, Mother Nature sets the Universe eventually hush-hush, under a limitation that there simply are a few mysteries that are Hers and Hers alone to know, and not for us simple humans. Be that as it may, truth be referred to, Mother Nature is similarly prohibitive on occasion in any event, when likelihood doesn't go into the condition. Consequently, quantum material science isn't some most essential thing in the world of neglecting to grapple with enormous assurances. Regardless, the idea of likelihood is human, and quantum material science originates before human thoughts. Quantum material science perhaps brimming with probabilities to us humans, yet not to Mother Nature.

Likelihood and quantum material science: the issue here isn't whether quantum physical science works - it's been demonstrated 100% exact down to the twelfth decimal spot, to say the very least. It is eventually liable for more than 1/third of the worldwide economy in mechanical thingamajigs and applications. The issue is that quantum material science plays the game and works under fixed and last guidelines of causality, or does it play by its own spontaneously 'rules' that aren't generally governed since they are intended to be broken.

Either causality works, or it doesn't. If it does, at that point, quantum material science doesn't, can't, swagger its stuff pell-mell with no circumstances, and logical results inactivity. If causality doesn't work, at that point, assurance doesn't work at any level since the sureness we partner with the full scale is based on the miniature's vulnerability.

Quantum vulnerability, or the contrary side of the coin, likelihood, usually is made express by the Heisenberg Uncertainty Principle, which essentially expresses that through no flaw of your own or your instrumentation, it is tough to know different differentiating properties about a critical molecule. The more you nail down and think around one property, the fuzzier another property becomes, and the other way around. You can never realize the two properties totally to a 100% sureness. Indeed you can never know either property to the 100% sureness level. That is because the very demonstration of watching or estimating changes the stuff you are attempting to watch or measure. Earth has constrained or put this not-to-be-arranged and no-correspondence-will-be-went into a limitation on you, the onlooker, or on your companion, your estimating thingamajig. So there! Or then again, is it so? The key is that you, the eyewitness, or your estimating doohickies gadget, is in a wicked way. You can't have the foggiest idea about the exact situation of the framework you are keen on if you are essential for that framework. You are not part of the arrangement; you are the issue!

The likelihood is simply an explanation that you, the human you, don't know something for supreme specific. That is it. When you discover for sure, it's no longer a likelihood except for conviction. On the off chance that you can't find, and the very demonstration of watching or estimating can adjust the properties of what you are attempting to watch or measure (and that is genuinely what the Heisenberg Uncertainty Principle is about), what happens or eventuates if there is no perception or estimation?

In each definition or clarification, I've ever

observed about the Heisenberg Uncertainty Principle, it is either inferred o expressly expressed that an onlooker or potential estimation is being endeavored or thought of.

The likelihood remains likelihood on the off chance that you can't know by and by or even in principle. Notwithstanding, one can hypothesize that an omniscient (all-knowing) God must know everything practically speaking and in code. No individual who has faith in an omniscient God could take any confidence in quantum material science as working in the domain of likelihood, same the Heisenberg Uncertainty Principle. Nonetheless, I genuinely don't have to go down that pathway since I state with conviction that there is no God, all-knowing, or something else.

Regardless of whether you don't have the foggiest idea, it is conceivable to know in principle, well, that it also results in any event hypothetical conviction.

In any case, imagine a scenario where it is absurd to expect to know, even in principle, a.k.a. the Heisenberg Uncertainty Principle? Indeed, that as well doesn't of need rule in likelihood and preclude assurance.

As another case of supposed quantum likelihood, take radioactive rot, which is affirmed to be missing, is causality - it occurs for reasons unknown by any stretch of the imagination. A radioactive particle, or its core, will rot, yet precisely when and under what conditions are unusual, perhaps in 10 seconds, possibly not for a billion years. It's all likelihood.

This is a case of Mother Nature concealing dearly held secrets. The spectator is upset in grappling with radioactive rot other than through, or by figuring probabilities. Consequently, quantum material science is a likelihood. However, that is just if you acknowledge the absence of causality premise. I reject that and propose that radioactive rot has a reason - we simply don't know what it is. Because of Mother Nature's wardrobe, we are confined or forestalled with absolutes or impediments to our vision of the real world. There are bunches of instances of skeletons in Mother Nature's storage room that don't include likelihood (see underneath), so for what reason should radioactive rot be a particular case to the standard?

If a human eyewitness is available, she may

state that the radioactive nuclear core has a 50-50 possibility of going poof in 60 minutes if dependent on registering probabilities. However, on the off chance that there is no human onlooker, the radioactive substance will go poof (sure beyond a shadow of a doubt) - in the end. There's no likelihood included because no fake time units had - time units are a human idea or innovation, not part of Mother Nature's jargon. So chance in quantum material science is spectator reliant (or subject to there being an onlooker) - no eyewitness, no possibility, just assurance.

The compelling force of nature has forced loads of different absolutes or restrictions on us. Hop into a Black Hole, and you're not coming out once more, regardless of whether you were conceived on Krypton. No likelihood here.

You can't go at the speed of light - period! No likelihood here.

If you are inside a shut room (no windows), you have no chance to tell if you are on Earth and in Earth's 1-G gravity field or space being quickened at 1-G. No likelihood here.

Like those mentioned above, you have no

feeling of movement while sitting efficiently on your couch. However, the Earth is turning on its hub; the Earth is circling the Sun; the Sun is circling the Milky Way Galaxy; and the Milky Way Galaxy and the Andromeda Galaxy are on an impact course (unwind, not to converge for another five billion years).

Similarly, if you were in a spaceship without any windows (no reasonable looking outside), and that spaceship was going at a steady pace, you wouldn't feel it, and in this way, you wouldn't know that you were going at a quick rate of bunches. No likelihood here.

The unstoppable force of life doesn't expect you to be brought forth; she expects you to pass on. No likelihood here.

You are on a train halted at the railroad station. To your left side is another train that additionally ended at the railroad station. That other train begins moving to your back, or are you pushing ahead deserting the other train. Which right? However, it's before long to be self-evident only for a couple of moments, and you didn't have the foggiest idea. On the off chance that all that existed were only the two trains and you with no different casings of reference, you'd be sure whether the other train was moving, or if your train was moving, or both. No likelihood here.

You can't watch any aspect of the Universe that lives into the great beyond that denotes the noticeable limit that contains the observable Universe (simply like you cannot watch a boat that has cruised into the great beyond the round Earth). No likelihood here.

At the point when you watch out into the night sky at the far off stars and worlds, you are thinking back in time, since it takes effort for the light of those items to contact us. Yet, you can't watch the Universe further ago than 300,000 years post that Big Bang occasion. That is because the Universe was still excessively thick with stuff to permit seeing. It's likened to how you can't see the focal point of the Sun because there's a lot of sun-stuff in the manner. It requires some investment for a photon to battle its way from the middle to the outside of the Sun. Along these lines, 300,000 years is the breaking point, which is the reason it's rubbish for cosmologists to direct with supreme sureness what the structure and substance of the Universe resembled preceding that time, mainly that garbage than a nanosecond after the Big Bang, the Universe was only the size of a pinhead - they are merely guesstimating and terrible guesstimating at that. No likelihood here.

You can't change the past. No likelihood here.

At long last, without our cutting edge innovation, the 'Stripped Ape' couldn't recognize gamma beams, or X-beams, or radio waves, or microwaves, astronomical rays, neutrinos, and a large group of different pieces and pieces that are an integral part of the Universe. No likelihood here.

So you see that Mother Nature has forced all way of supreme snags in our method of looking into her skirt and revealing her 'private' nature. That doesn't mean the life systems don't exist, just we're not permitted to look, and there's not a damn thing we can do about it. Thus, her life structures are unsure or presumably is this, or that or the following thing yet just to us, the wannabe eyewitness.

Finally, consider and reexamine the quantum mantra: Anything that isn't prohibited is obligatory; anything that can happen will occur.

Does that sound like a likely explanation for you?

In synopsis, all in all, references to quantum material science are loaded with "likelihood." They are additionally loaded up with terms connecting likelihood to somebody like me or somebody like you - an eyewitness. Eliminate or take out the spectator, and you eliminate or kill the possibility in quantum probability.

Quantum mechanics is the part of material science identifying with the tiny.

It brings about what may have all the earmarks of being some abnormal decisions about the physical world. At the size of particles and electrons, many traditional mechanics conditions, which depict how things move at regular sizes and speeds, stop to be valuable. In old-style mechanics, objects exist in a particular spot at a specific time. In any case, in quantum mechanics, objects rather live in a fog of likelihood; they have a distinct possibility of being at point A, one more opportunity of being at point B, etc.

**Three progressive standards**

Quantum mechanics (QM) was created over numerous years, starting as many dubious numerical clarifications of investigations that traditional mechanics couldn't math clarify. It began at the turn of the twentieth century, around a similar time that Albert Einstein distributed his physics of relativity, a different numerical transformation in material science that depicts the movement of things at high speeds. In contrast to relativity, however, QM's birthplaces can't be ascribed to anyone's researcher. Or maybe, numerous researchers added to an establishment of three progressive rules that bit by bit picked up acknowledgment and experimental confirmation somewhere in the range of 1900 and 1930. They are:

Quantized properties: Certain properties, for example, position, speed, and shading, can now and again just happen in explicit, set sums, much like a dial that "clicks" from number to number. This tested a basic supposition of old-style mechanics, which said that such properties should exist on a smooth, constant range. To portray the possibility that a few properties "clicked" like a dial with explicit settings, researchers authored "quantized."

Particles of light: Light can now and then carry

on as a molecule. At first, this was met with cruel analysis, as it negated 200 years of examinations indicating that light acted like a wave, much like waves on the outside of a quiet lake. Light carries on also in that it ricochets off dividers and curves around corners, and that the peaks and box of the wave can include or counterbalance. Formed wave peaks bring about more glorious light, while waves that cancel produce dimness. A light source can be thought of as a ball on a stick being musically plunged in the focal point of a lake. The shading discharged compares to the separation between the peaks, which is dictated by the speed of the ball's mood.

Floods of issue: Matter can likewise carry on like a wave. This opposed the approximately 30 years of analyses demonstrating that issue (for example, electrons) exists as particles.

**Quantized properties?**

In 1900, German physicist Max Planck tried to clarify hues' circulation over the range in the sparkle of scorching and white-hot articles, for example, light fibers. When comprehending the condition he had inferred to portray this dispersion, Planck acknowledged it suggested that blends of just certain hues (yet an incredible number of them) were transmitted, explicitly those that were exclusive number products of some base worth. Some way or another, hues were quantized! This was unforeseen because the light was perceived to go about as a wave, implying that shading estimations should be a nonstop range. What could be denying molecules from delivering the hues between these entire number products? This appeared to be peculiar to the point that Planck viewed quantization as merely a numerical stunt. As indicated by Helge Kragh in his 2000 article in Physics World magazine, "Max Planck, the Reluctant Revolutionary," "If a revolution happened in physics in December 1900, no one appeared to see it. Planck was no exemption …"

Planck's condition likewise contained a number that would later turn out to be critical to QM's future improvement; today, it's known as "Planck's Constant."

Quantization assisted with clarifying different riddles of material science. In 1907, Einstein utilized Planck's speculation of quantization to explain why the temperature of a strong changed by various sums on the off chance that you put a similar measure of warmth into the material yet changed the beginning temperature.

Since the mid-1800s, the study of spectroscopy had indicated that various components radiate and retain exact shades of light called "otherworldly lines." Though spectroscopy was a dependable technique for deciding the features in articles, for example, inaccessible stars, researchers were perplexed concerning why every part emitted those particular lines in any case. In 1888, Johannes Rydberg determined a condition that depicted the unearthly lines transmitted by hydrogen; however, no one could clarify why the state worked. This changed in 1913 when Niels Bohr applied Planck's speculation of quantization to Ernest Rutherford's 1911 "planetary" model of the molecule, which proposed that electrons circled the core a similar way that planets orbit the sun. As per Physics 2000 (a site from the University of Colorado), Bohr recommended that electrons were limited to "uncommon" circles around a particle's core. They could "hop" between exceptional rings, and the energy created by the hop caused vivid shades of light, seen as ghastly lines. Even though quantized properties were developed as yet a simple numerical stunt, they clarified so much that they turned into the establishing standard of QM.

**Particles of light?**

In 1905, Einstein distributed a paper, "Concerning a Heuristic Point of View Toward the Emission and Transformation of Light," wherein he imagined light voyaging not as a wave, however, as some way of "energy quanta." This bundle of energy, Einstein proposed, could "be retained or produced distinctly overall," explicitly when a molecule "hops" between quantized vibration rates. This would likewise apply, as indicated by a couple of years after an electron "bounces" between quantized circles. Under this model, Einstein's "energy quanta" contained the hop's energy distinction; when isolated by Planck's consistent, that energy contrast decided the shade of light conveyed by those quanta.

With this better approach to imagine light, Einstein offered bits of knowledge into the conduct of nine unique marvels, including the particular hues that Planck depicted being discharged from a light fiber. It likewise clarified how certain shades of light could launch electrons off metal surfaces, a wonder known as the "photoelectric impact." However, Einstein wasn't completely defended in taking this jump, said Stephen Klassen, a material science partner at the University of Winnipeg. In a 2008 paper, "The Photoelectric Effect: Rehabilitating the Story for the Physics Classroom," Klassen states that Einstein's energy quanta aren't crucial for clarifying those nine marvels. Certain numerical medicines of light as a wave are as yet equipped for depicting both the particular hues that Planck portrayed being produced from a light fiber and the photoelectric impact. For sure, in Einstein's disputable winning of the 1921 Nobel Prize, the Nobel board just recognized "his revelation of the law of the photoelectric impact," which explicitly didn't depend on the thought of energy quanta.

Approximately twenty years after Einstein's paper, the expression "photon" was promoted for portraying energy quanta, on account of the 1923 work of Arthur Compton. He demonstrated that light dispersed by an electron bar changed in shading. This showed particles of light (photons) were undoubtedly crashing into particles of issue (electrons), hence affirming Einstein's speculation. At this point, light could act both as a wave and a molecule, setting light's "wave-molecule duality" into the establishment of QM.

www.ingramcontent.com/pod-product-compliance
Lightning Source LLC
Chambersburg PA
CBHW070638220526
45466CB00001B/225